Table of Contents

Introduction

"If the pictures of tens of thousands of human bodies being gnawed on by dogs do not wake us out of our apathy, I do not know what will."[1] Nobel Peace Prize laureate and Secretary-General of the United Nations Kofi Annan made this emotional statement regarding the 1994 genocide that occurred in Rwanda. In that country, genocide killed over 800,000 innocent men, women, and children in 100 days.[2] As the genocide occurred, the international community waited and watched the horrific events unfold. The catchphrase, "never again" after the Holocaust of World War II became once again an empty promise.

Genocides and mass atrocities evoke individual emotional responses and empathy, but did not necessarily affect national or international interests. Hence, the international community did not take any practical measures to prevent or stop the occurrence of genocide. However, at the turn of the millennium, a movement was afoot in the world to change their policies. In 2005, a United Nations report laid the foundation for the world community to invoke the responsibility to protect (R2P) people inflicted by mass atrocity and genocide.[3] This initiative aimed to limit the occurrence of mass atrocity and genocide throughout the world. The United States followed suit in 2006 with its National Security Strategy, identifying genocide as intolerable.[4]

This shift in U.S. policy provided the momentum for the Carr Center for Human Rights Policy and the Peacekeeping and Stability Operations Institute (PKSOI) of the U.S. Army War College to partner. This partnership established the Mass Atrocity Response Operations (MARO) project.[5] This project's goals are to create a widely shared understanding of mass atrocities and

---

[1] Kofi Annan, "Undersecretary-General of the United Nations," *The East African* (March 1996).
[2] Romeo A. Dallaire, *Shake Hands with the Devil: The Failure of Humanity in Rwanda* (New York, NY: Carroll & Graf Publishers, 2003), xvii.
[3] United Nations General Assembly, *2005 World Summit Outcome*, 60th sess., September 15, 2005, 31-32.
[4] George W. Bush, *The National Security Strategy of the United States of America* (Washington, DC: The White House, March 2006), 17.
[5] Sarah Sewall, Dwight Raymond, and Sally Chin, *Mass Atrocity Response Operations: A Military Planning Handbook* (Charleston: Createspace, 2010), 5.

to create a military approach to resolve them. One product of the MARO project was a military planning handbook, designed to be a planning guideline for military operations in MARO. The handbook's scope focuses on joint operational planning, and is thus, limited in nature. It does not cover in detail the role of air power in MARO.

The purpose of this paper seeks to answer the question, how does the United States Air Force best execute its role in MARO? **The analysis will reveal practical air power solutions to this problem.** The intention of this paper is not to be an addendum to the MARO handbook; but to expand on the foundation of knowledge provided by it. To examine the hypothesis, the paper will assess the efficacy by using prescriptions from Colonel John Boyd's *A Discourse on Winning and Losing*. Colonel Boyd's work on tactical and strategic isolation offers insights into the effects air power can have on the enemy, or the perpetrator-groups of mass atrocity. Next, the paper examines the current literature relative to MARO and provides key foundations and principles for combating mass atrocities. These principles provide the context for general rules and strategies in MARO. The paper then provides an analysis of current Air Force doctrine to provide salient air power solutions used in conjunction with other instruments of power to combat MARO. The paper concludes with recommendations ranging from military preparation to doctrinal changes.

Literature Review

Perhaps the greatest contributor to the prevention and punishment of genocide would be Raphael Lemkin. In fact, he coined the term genocide by combining the Greek word *geno*, meaning "race" or "tribe", with the Latin word *cide*, meaning "killing."[6] His life's work to prevent genocide and establish international law to punish it culminated in 1948 when the United Nations drafted the Convention on the Prevention and Punishment of the Crime of Genocide. Article II of the Convention defines genocide as:

---

[6] Samantha Power, *A Problem from Hell: America and the Age of Genocide* (New York, NY: Basic Books, 2002), 42.

In the present Convention, genocide means any of the following acts committed with intent to destroy, in whole or in part, a national, ethnical, racial or religious group, as such:

- Killing members of the group;
- Causing serious bodily or mental harm to members of the group;
- Deliberately inflicting on the group conditions of life calculated to bring about its physical destruction in whole or in part;
- Imposing measures intended to prevent births within the group;
- Forcibly transferring children of the group to another group.[7]

Despite the ratification in 1951 of the U.N. Convention on the Prevention and Punishment of the Crime of Genocide, it was not until 1998 that the first case of genocide appeared in an international criminal tribunal.[8] This 47-year void of action in the prevention or punishment of genocide was notably due to the issue of national self-interest and an emphasis on state sovereignty.

Samantha Power's book, "A Problem from Hell: America and the Age of Genocide," is the most persuasive, comprehensive study of genocide, ranging from Raphael Lemkin's work to post 9/11 sentiment in America. Her work chronicles the Armenian Genocide in 1915, through Cambodia, Iraq, Bosnia, Rwanda, Srebrenica, and Kosovo. It not only details the accounts of the actual perpetrator-groups of the genocide, but also the legal ramifications within the international community and the United States. It concludes with a brief description of her expected reaction of the public after the September 11, 2001 attacks in the United States. She proposes the United States should stop genocide for two reasons. The first reason is the moral implication of genocide and responsibility to protect human life. The second reason, notably more practical when viewed from national self-interests, is international stability. After 9/11, and the global war on terror, it was in the U.S. self-interest, to ensure genocide did not occur. Genocide, and the regional instability it creates, is not conducive to U.S. security needs.[9]

---

[7] United Nations General Assembly, *Convention on the Prevention and Punishment of the Crime of Genocide,* Adopted by Resolution 260, December 9, 1948, III (A).

[8] Samantha Power, *A Problem from Hell: America and the Age of Genocide,* 484.

[9] Ibid., 511-516.

To answer the question of state sovereignty or non-interference, UN Secretary-General Kofi Annan posed the question: "if humanitarian intervention is, indeed, an unacceptable assault on sovereignty, how should we respond to a Rwanda, to a Srebrenica—to gross and systematic violations of human rights that offend every precept of our common humanity?"[10] The International Commission on Intervention and State Sovereignty (ICISS) formed to answer this question. After intensive research and deliberation, the Commission's report, "The Responsibility to Protect" addressed the state sovereignty issue. This report's scope covered critical issues including: authority to intervene, legal implications of intervention, decision criteria, and military action. The main theme from this report concludes that nation-states have the responsibility and rights to protect their own citizens. However, if nation-states are unable, or more importantly unwilling to protect their citizens, the responsibility to protect those citizens is the responsibility of the international community.[11] This report was the genesis of the responsibility to protect (R2P), an ongoing legal and moral precedence.

The next evolution in the international community to implement R2P was the "2005 World Summit Outcome" document. The heads of state and government on the United Nations General Assembly agreed to its outcome. Although there were many different agendas of this meeting, it included the validity of R2P. The General Assembly concurred that it was the nation-states responsibility to protect its own people. However, "we are prepared to take collective action, in a timely and decisive manner ... should peaceful means be inadequate and national authorities manifestly fail to protect their populations from genocide, war crimes, ethnic cleansing and crimes against humanity."[12] This document was the first formal document permitting external military action against genocide if sovereign means were unattainable.

---

[10] Kofi A. Annan, *The Millennium Report of the United Nations Secretary-General, 2000* (New York, NY: United Nations, Department of Public Information, 2000), 48.

[11] Gareth Evans and Mohamed Sahnoun, *The Responsibility to Protect.* (Ottowa, CA: International Development Research Centre, 2001), 6.

[12] United Nations General Assembly, *2005 World Summit Outcome*, 60th sess., September 15, 2005, 31-32.

The United States path toward genocide prevention and R2P followed that of the international community. In 2006, President Bush stated in the National Security Strategy his vision of genocide prevention. The United States would no longer tolerate acts of genocide. He noted that American inaction, as the sole superpower, led to many other nations deciding not to act. The U.S. should implement instruments of national power: diplomatic, informational, military, and economic as required in mass atrocity events. If peaceful intervention was not possible, armed intervention may be required.[13] The adoption of genocide prevention by the President paved the way for other departments within the government to take proactive steps to implement this strategy. Two separate efforts followed the 2006 NSS, the creation of the Genocide Prevention Task Force for U.S. policy-makers and the Mass Atrocity Response Operations Project for the U.S. military.

Madeleine Albright and William Cohen were the co-chairs of the Genocide Prevention Task Force. In 2008, they published, "Preventing Genocide: A Blueprint for U.S. Policymakers". This work took genocide prevention out of the realm of ideology and into that of reality. The main effort of this work was to demonstrate a need for a comprehensive policy for U.S. leaders, in order to implement a cohesive capability to respond to genocide. The first portion of this book focused on leadership within the government and how to create a political will to push for prevention. It also went as far to recommend assigning 250 million dollars annually to this project.[14] The strategy highlighted the need for: policy allowing for early prevention of genocide, diplomatic measures to stop genocide, coalition or international partners, and the possibility for military action as a last resort.

Sarah Sewell founded the U.S. MARO project in 2007. Its efforts to build a common military approach to preventing and stopping mass atrocities led to the creation of a military

---

[13] George W. Bush, *The National Security Strategy of the United States of America* (Washington, DC: The White House, March 2006), 17.

[14] Madeleine K. Albright and William S. Cohen, *Preventing Genocide: A Blueprint for U.S. Policymakers* (Washington, DC: Genocide Prevention Task Force, United States Holocaust Memorial Museum, 2008), xvii.

planning handbook." This handbook's intention is to prepare the military operationally for the possibility of genocide intervention, not necessarily to advocate for its use.[15] The MARO handbook developed a conceptual framework for military options, by comparing it to other types of options. It also provides a series of flexible deterrent options the National Command Authority could implement in the event of military intervention.

Beyond the auspices of the MARO project, other military literature is readily available. One particular article of importance to this paper is the work of Douglas Peifer. His article, "Genocide and Air Power" provides a set of air power capabilities applicable to stopping genocide. These include strategic airlift, communications support, medevac support, radio suppression and broadcast capability, and offensive air power capabilities.[16] While not an inclusive list of capabilities this paper advocates, there is overlap between his work and that of this paper.

One difficulty in combating mass atrocities and genocide is isolating the perpetrator-groups. This isolation must obviously include physical isolation, separating the perpetrator-groups from the victims to stop the act of genocide. However, there must also be a mental and moral isolation of the perpetrator-groups to stop the acts of genocide.[17] Colonel John Boyd's theoretical work on isolation of enemies is the perfect medium for researching this concept with respect to MARO. Frans P.B. Osinga's book, "Science, Strategy and War: The Strategic Theory of John Boyd," provides a comprehensive study of Col. Boyd's work, since most of Boyd's work consisted of a series of presentations, with relatively few short essays.

To date, the evolution of R2P and the influence it has on the United States continues. President Obama has continued the legacy started by his predecessor. The 2010 United States

---

[15] Sarah Sewall, Dwight Raymond, and Sally Chin, *Mass Atrocity Response Operations: A Military Planning Handbook*, 12.

[16] Douglas Peifer, "Genocide and Air Power." *Strategic Studies Quarterly* 2, no. 2 (Summer 2008): 112-119.

[17] Frans P.B. Osinga, *Science, Strategy and War: The Strategic Theory of John Boyd* (New York, NY: Routledge, 2007), 209-219.

National Security Strategy states, "The United States is committed to working with our allies, and to strengthening our own internal capabilities, in order to ensure that the United States and the international community are proactively engaged in a strategic effort to prevent mass atrocities and genocide."[18]

Methodology

The analysis of this paper will follow the grounded practical theory in its methodology.[19] As such, it will provide a normative theory to guide the practical application of air power to MARO. The first step in the analysis is a comprehensive study of Colonel John Boyd's work on enemy isolation. This analysis will be both inductive and deductive, detailing vulnerabilities in the perpetrators of mass atrocity. This will provide the "problem level" of the grounded practical theory.[20] The second step in the analysis will be to review relevant literature and doctrine. It will first examine the "Mass Atrocity Response Operations: a Military Planning Handbook." This handbook conforms to the U.S. military's Joint Operational Planning Process and identifies a series of obstacles to MARO.[21] Next, it will examine relevant joint doctrine in order to scope the paper. The integration of the perpetrator-groups vulnerabilities provided by Boyd's work and integration of the MARO handbook analysis will reveal several niches in which air power is perfectly suited to fill. This will provide the "technical level" of the grounded practical theory.[22] Finally, the paper will examine Air Force doctrine to provide salient solutions and air power standards to shape the practice and operational art of MARO. This will provide the "philosophical level" of the theory.[23]

---

[18] Barack H. Obama, *The National Security Strategy of the United States of America* (Washington, DC: The White House, May 2010), 48.

[19] Robert T. Craig and Karen Tracy. "Grounded Practical Theory: The Case of Intellectual Discussion." *Communication Theory* 5, issue 3 (March, 1995): 248-272.

[20] Ibid., 253.

[21] [21] Sarah Sewall, Dwight Raymond, and Sally Chin, *Mass Atrocity Response Operations: A Military Planning Handbook*, 41.

[22] Craig, Tracy, "Grounded Practical Theory: The Case of Intellectual Discussion" 253.

[23] Ibid.

Isolation of the Enemy

Over the last decade, the United States has made substantial progress in the development of policies to combat mass atrocities. Many of these initiatives focus on preventing mass atrocities vice intervening once mass atrocities or genocides begin. There is no doubt prevention is preferred, namely for its innate ability to protect more human life. Prevention is also favored because it is a less complicated problem to solve. The cohesion of future perpetrator-groups before acts of genocide is more susceptible to intervention and is accessible to a myriad of international organizations.[24] Once hostilities commence however, access is limited to very few organizations, namely military or police organizations able to protect themselves from the violence that is taking place. Perpetrator-groups involved in the acts of genocide have also evolved into self-regulating, hierarchical organizations less susceptible to intervention. Finally, once genocides begin, it is much more difficult to isolate perpetrator-groups, or to isolate the enemy. Although it is more difficult to stop genocides once they have begun, it is evident in recent policy that the United States is undertaking this initiative as well.

Colonel John Boyd's theoretical work on isolation of the enemy is a perfect starting point for the analysis of this paper. Colonel Boyd, a United States Air Force fighter pilot, is well known for his Observation-Orientation-Decision-Action loop, referred to as the OODA loop. The OODA loop is most commonly understood as a rapid decision making process, allowing you to out-think and out-pace your opponent.[25] However, the OODA loop is the culmination of a more substantial body of work, *A Discourse on Winning and Losing.* Although there are many practical uses for his work, Colonel Boyd evidently also wanted to educate his audience in the construction of his theory. Colonel Boyd worked on this series of presentations starting in 1976 and ended in

---

[24] For the purpose of this paper, the definition of perpetrator-groups is a group of people preparing to commit genocide, willing to commit genocide, or is committing genocide.

[25] Frans P.B. Osinga, *Science, Strategy and War: The strategic theory of John Boyd,* 1.

8

1995, two years before his death.[26] This paper will focus primarily on his work in isolating the enemy.

> In order to defeat an enemy, Boyd suggests our strategy should encompass the following:

> Penetrate adversary's moral-mental-physical being to dissolve his moral fiber, disorient his mental images, disrupt his operations, and overload his system, as well as subvert, shatter, seize or otherwise subdue those moral-mental-physical bastions, connections, or activities that he depends upon, in order to destroy internal harmony, produce paralysis, and collapse adversary's will to resist.[27]

This statement alone does not fully encompass Boyd's ideas, but it is the primary source of the analysis in this section. Before the author can begin to dissect the implications of this, the reader must understand a foundation of his post-modernistic theory.

Foundation of Boyd Theory

Boyd argues in his work that an adversary, in our case the perpetrator-group, is an open system. Within the adversary, there are a multitude of individuals and hierarchal structures unifying them. All of the individuals or sub-systems connect to each other by linkages, whether they are communication methods, social relationships, command relationships, cultural ties, or ideological values. The important part is these linkages create an interconnected system. These linkages are not only open to inputs from other members of the system, but are also open to outside sources. In the case of the perpetrator-group, local society and cultural norms are two examples of outside sources of input. Since the adversary system is open to inputs from outside itself, it is an open-system.[28] Thus, an open-system interacts with its environment, not only social and cultural contributions, but also those of an adversary.

Since open-systems interact with an outside environment, they are intrinsically dynamic. Outside inputs consists of an almost infinite number of variables, constantly changing the parts and linkages of the interconnected system. Boyd defines the range of interactions within this

---

[26] Ibid., 1-19.
[27] Ibid., 176.
[28] Ibid., 88.

9

system as the cognitive domain.[29] As the system becomes more advanced and hierarchal, the cognitive domain must develop in order to cope with these changes. Thus, there is a degree of complexity within the system. In order for the system to survive, it must continually adapt. This adaptation tends toward emergence, where interaction in the system leads to markedly different behavior patterns than those of the individual components.[30] Put simply, the system evolves and from this evolution, a new code of conduct emerges, different from the parts within the system. For example, in the genocide perpetrator-group, the interactions and complexity within the system cannot cope with changes. The system must adapt to these variables, and as it does, the emergence of genocidal tendencies occurs to strengthen the system. Although the emergence of genocide within the system may be conducive to the organizational structure's viability, it is not instinctive for individual people within the group to commit genocide.

James Waller examines aspects of this process in his book, "Becoming Evil: How Ordinary People Commit Genocide and Mass Killing."[31] As an organization changes and genocide emerges, the system must develop means to enable the system to commit genocide. Waller breaks this into three constructs: the "cultural construction of worldview", the "psychological construction of the other", and "social construction of cruelty."[32] The first examines cultural models of the perpetrator-group that are widely shared among the members. The "psychological construction of the other" describes how perpetrators turn the victims of genocide into 'objects' of their actions, despite the individual not being homicidal in nature. The final construct deals with the coping process of the perpetrators to deal with their actions.[33] These three constructs are linkages within the system and are thus open to outside influence.

---

[29] Ibid., 94.
[30] Ibid., 95.
[31] Dr. James Waller is Professor and Cohen Chair of Holocaust and Genocide Studies at Keene State College. He is a widely renowned scholar in genocide studies.
[32] James Waller, *Becoming Evil: How Ordinary People Commit Genocide and Mass Killing* (Oxford: University Press, 2002), 9-12.
[33] Ibid., 12.

In order to deal with the complexity inherent within any developed system, Boyd further theorizes any open-system must seek to be an adaptive self-organization.[34] Within the adversary's system, there are many developed relationships between entities. These linkages are the fiber that not only hold an organization together, but also expose it to intervention. To truly exploit these linkages, an understanding of how they connect is paramount. Boyd argues there is a two-fold process to begin this understanding process. The first of these methods is deduction, breaking apart an entity from general to specific.[35] A deductive process starts from a complete whole and breaks it down into individual parts. For example, breaking a perpetrator-group down to its individual elements: ideology, cultural norms, command hierarchy, and leadership. The second process to have a complete understanding of linkages is induction, or synthesizing individual parts into a complete whole.[36] In our example, it would consist of examining habit patterns, religious affiliation, and financial hardships in order to integrate them into a whole. While both methods are independently viable, Boyd argues that a cyclical pattern of deduction, induction, deduction, etc. is required to fully appreciate and understand a system.[37]

The final piece of Boyd's post-modernistic foundation applicable to this paper is the concept of deconstructionism. Deconstructionists argue people view the world and its infinite systems through the lens of their own perspective. They cannot view the world objectively because of their own preconceptions: history, prejudices, ideology, language, and social status.[38] Thus, Boyd theorizes learning occurs from social interchanges between people. This simple statement is critical in the understanding of the adversary. The perpetrator-group has its own values and hierarchy, which in effect is its reality. In order to understand it, there must be some degree a social exchange with the system. The ability to analyze and synthesize the parts with in

---

[34] Frans P.B. Osinga, *Science, Strategy and War: The strategic theory of John Boyd,* **88.**
[35] Ibid., 133.
[36] Ibid.
[37] Ibid., 134.
[38] Ibid., 110.

the social context of the system must occur, another argument for the deduction and induction process.

Practicality of Boyd Theory

The foundation of Boyd's work allows the reader to fully engage with the cognitive process required to isolate the enemy. The goal is to, "Collapse adversary's system into confusion and disorder by causing him to over and under react to activity that appears simultaneously menacing as well as ambiguous, chaotic, or misleading."[39] Building on the foundation provided in the last section, planners must develop methods to stop the adversary from being able to quickly adapt to the continuum of changes. Thus, they must first fully understand the enemy's system, its parts and linkages, through the process of induction and deduction. Once the planner understands the system, they can identify the weaknesses and vulnerabilities. This will provide the solution to create isolation of the enemy, through confusion and disorder.

Boyd identified several methods planners can use to create this confusion and disorder within the system. These abstract strategies are useful against any open-system, adaptive organization. The first path already identified is to have a complete understanding of the enemy system. Planners must identify the parts and linkages of the enemy system, but they must also be fully aware of how the friendly and enemy systems connect to each other. Once there is an understanding of the system, planners must identify multiple strategies or avenues of approach to cause disorder within the enemy system.[40] Boyd argues multiple strategies are required because any complex system will have redundant linkages. In order to cause the most confusion and disorder, simultaneity of multiple attacks against an enemy system will directly affect primary linkages but also their redundant ones.

---

[39] Ibid., 141.
[40] Ibid., 126

12

Following the same thought process, Boyd highlights the need to overload the cognitive capacity of the enemy system.[41] Using simultaneity once again, planners must attack critical cognitive nodes or linkages within the system. In the perpetrator-group example, command and control elements would be the cognitive nodes, the leaders of the group. Attacking the cognitive nodes and the elements of communication, the cognitive linkages, would critically hamper the viability of the system. He further argues to negate or eliminate all crucial subsystems within the system. This will diminish the opportunity of the system to react to changes in its environment.

To fully realize the effects highlighted above, planners must negate the elements of the physical and social environment that support the system.[42] This could be as simple as removing the economic viability of the system, or as complicated as severing the support of the local populace. In the latter case, cutting the information flow from the populace, the social environment, and the system may critically hamper survivability. In general, planners must attack any support to the system. Close off any support and amplify any resistance to it.

The final piece Boyd addresses in order to cause confusion to the system is to attack the schemata or plans of the enemy system.[43] The enemy system is not going to idly sit as changes to its environment and cognitive nodes occur. Thus, planners must negate any reactions of the system, or at a minimum, stop the ability to validate the reactions. The enemy system might react correctly, but if it does not know it is making the correct reaction, more disorder and confusion may occur. This can include the nature of the conflict. By changing the nature of the conflict, the form of warfare may be outside the understanding of the enemy system. This can include changes to the alliances with other systems, the location of the conflict, or even the stakes involved.

Taking the lessons learned from Boyd's theory, the paper will now shift from abstractness to concrete ways to isolate those perpetrator-groups committing mass atrocities.

---

[41] Ibid.
[42] Ibid.
[43] Ibid.

There are three categories Boyd identifies to isolate the enemy: physical isolation, mental isolation, and moral isolation. Physical isolation occurs by cutting both internal and external communication within the system. Mental isolation occurs by ambiguous and deceptive inputs to the system and operating at a faster operations tempo than the enemy can sustain. The enemy morally isolate themselves by their own actions, violating accepted codes of conduct.[44] This paper will use these three elements to categorize specific possibilities to combat mass atrocities.

Physical Isolation

The perpetrator-group as an enemy system is as vulnerable as other military organizations, if not more so. In the spirit of Boyd's theory, planners must attempt to attack all of these vulnerabilities simultaneously. Quite possibly the most critical node in the perpetrator-group system is the leadership. Whether a state or non-state actor, the leadership generally provides the impetus for the acts of genocide. It also provides the input to the command and control function to the entire system. Thus, planners should attack this node of the enemy system from the onset. Removal of the leadership is the ultimate goal, either killing or capturing it. However, due to its inherent criticality to the system, it will be very hard to locate and destroy. Despite this, the linkages to this critical node are susceptible at the start. Disrupting communication between the leadership node and other nodes of the system will effectively remove command and control.

Communication between the leadership and the other parts of the system is crucial, but to truly isolate the system from its environment, disruption of all communication within the system is required. The system's means to mass communication provides the highest payoff. In particular, planners should silence all forms of television and mass radio broadcasts endemic to supporting genocide. This would not only limit propaganda supporting genocide, but also remove recruiting methods of the perpetrator-groups. At the micro-level, targeting radio, phone, and Internet communication between factions of the perpetrator-group will also be effective at

---

[44] Ibid., 214.

14

limiting command and control. By removing major electronic forms of communication, the perpetrator-group is limited to face-to-face communication and couriers. Therefore, the next approach to limit communication is to remove or restrict movement within the area of operations. Restricting freedom of movement in the area of operations of the enemy hampers face-to-face communication.

Isolating key infrastructure or choke points within the enemy system critically hampers enemy operations for other reasons as well. Enemy logistics is required to sustain operations in genocide, and the backbone of logistics is transportation and infrastructure. Effectively controlling the infrastructure also limits the mobility of the perpetrator-group. If the perpetrator-group is unable to get access to the victims, genocide can no longer occur.

As mentioned, the perpetrator-group requires logistics to sustain its operations. There are multiple approaches available to the planner to limit the effectiveness of logistics. One approach would be to eliminate the economic means of the enemy. The removal of alliances at the strategic level could end financial support of the perpetrator-group. Seizure of current funds would also severely hamper operations. However, many perpetrator-groups may already be financially independent. In this case, planners must analyze sources of funding and remove them if able. The next approach to limit the effectiveness of logistics would be to remove the availability to distribute them. As mentioned, isolating the availability of transportation and the area's infrastructure are crucial. The last means to affect logistics would be to remove the actual supplies, or the ability to create the supplies. Trade embargoes or physically containing the area of operations would cut the enemy from the environment. Targeting the means to organically produce the logistics may also be a solution.

Physically isolating the perpetrators of genocide may not be attainable or it may take too long to occur. However, genocide can no longer occur if you remove the victims from the equation. While this might be unattainable in some situations, it warrants a place in this paper. Physically isolating the victims from the perpetrator-groups could be as simple as controlling the

15

infrastructure and means of transportation of the enemy, not allowing them to gain access to the victims. This obviously depends on the geographic constraints of the environment as well as how inter-mingled the perpetrators and victims are. If not feasible to separate them, planners could use armed forces to provide a zone of safety for the victims by physically protecting the victims. The last option would be to evacuate the victims from the area of operations. While not necessarily a permanent option, it also would stop the genocide from occurring.

The last option to physically isolate the enemy is to kill or capture the actual perpetrators. This would have a two-fold effect. First, it would remove perpetrators and reduce the amount of violence occurring. The second effect would contribute to the mental isolation of the enemy by raising the stakes of the conflict.

Mental Isolation

Mental isolation occurs when a planner uses deceptive and ambiguous inputs to the system to cause disorder and confusion. Boyd's conceptual work on the OODA loop plays a vital role in mental isolation, by disorienting the enemy so they can no longer understand what is going on.[45] This allows the planner the ability to out pace the enemy's operations tempo. To gain a faster operations tempo, planners must also create an operation with a structure to be mentally agile and adaptive. While many of the elements of physical isolation have effects here, there are several key additions within this category.

The first method to cause disorder and confusion within the system is to limit all forms of enemy communication. As mentioned above, this would physically limit the ability for the system to operate at full capacity. More importantly, this would decentralize the perpetrators of mass atrocity from the group. The perpetrators would then not be able to take orders or receive intelligence about the operation. Without the control from above, perpetrators would be confused on actions to take, making them susceptible to a local narrative.

---

[45] Ibid., 214.

16

Introduction of a local narrative is the next step in mental isolation. As the enemy system is reacting to confusion and disorder, it is more receptive to outside inputs as it attempts to adapt. To fully utilize simultaneity, planners can progressively input messages to underscore genocide while leveraging consequences for negative actions. The emergence of genocide might be conducive to the enemy system, but is not instinctive in the individual. By negating enemy communication, the individual is now dependent on his or her own moral compass. Planners should introduce a local narrative through the now open communication nodes, focusing on the individuals vice the enemy system. Social constructs in the enemy system have changed to make genocide acceptable in the minds of the individuals, thus changing the social constructs to make it unacceptable is required. The local narrative should be a multi-dimensional account of the effects of genocide. Examples of these would be descriptions on the negative effects of genocide on the economy. Local religious values would also be a positive step to pacify the perpetrators. The local narrative should also include the negative consequences for the perpetrator once captured.

Individuals within the enemy system feel safe from retribution for their actions. Thus, the negative consequences for their actions must be certain. Planners can exploit this in the local narrative by explaining the consequences for acts of genocide. This narrative should include examples of justice. The capture, trial, and punishment of other perpetrators in the enemy system are perfect for this. By invoking fear in the individual, he or she should be less willing to commit genocide.

Replacing peer pressure or perpetrator-group pressure, with social pressure, can further exacerbate this lack of impunity. Social pressure can come from the international community, local community, or an individual's family. Many times, those perpetrators committing genocide are not subject to an outside observer. Thus, there is not fear of any retribution or justice. Thus, planners should implement all available methods to collect intelligence on the actual perpetrators. Once collected, sharing this intelligence with society is instrumental. Even if not apprehended, the perpetrators are now subject to scrutiny from others within the community, which may have a stronger effect than capturing them.

Moral Isolation

Moral isolation occurs when the perpetrator-group violates established codes of conduct. The United Nations clearly defined genocide as, "acts committed with intent to destroy, in whole or in part, a national, ethnical, racial or religious group."[46] If perpetrator-groups violate this established code of conduct, then they should be subject to it. Many times throughout recent history, this has not been the case. Thus, to further isolate the enemy, planners should include methods to release information about the genocide to the international community. The more the international community knows about the atrocities being committed, the more willing they are to act on it. Even if nation-states are not willing to commit forces to combat genocide, international pressure to establish trade embargoes and the like, will hamper the effectiveness of the perpetrator-group.

Mass Atrocity Response Operations

The analysis provided in the first section of this paper provides measures to isolate the perpetrators of mass atrocities. Although practical measures, there are a myriad of operational implications when combating mass atrocities. In order to provide structure to the analysis provided in the previous section, this paper will examine current military literature provided in Joint Publications and the MARO handbook. These works will provide a foundation of operational planning, integral in providing relevant air power solutions when combating mass atrocities.

---

[46] United Nations General Assembly, *Convention on the Prevention and Punishment of the Crime of Genocide,* Adopted by Resolution 260, December 9, 1948, III (A).

Foundation of MARO Military Planning Handbook

The MARO Military Planning Handbook's goals are to provide planners fundamental characteristics they must understand, and explain the operational implications involved.[47] The handbook details three main distinctions between MARO and traditional warfare: multiparty dynamics, illusion of impartiality, and escalatory dynamic.[48] All of these distinctions invariably change the operational framework for which planners must contend. In MARO situations, a multiparty dynamic is fundamentally different from the traditional friendly versus enemy dynamic. There are multiple players in MARO situations providing even more difficulties when trying to isolate the perpetrator. The players in MARO are the perpetrator-group, the victims, the intervener, and other outside agencies.[49] As the intervener, the military group must ascertain the many nuances between the perpetrator-group and the victims while being impartial.

The concept of impartiality seems to contradict the effects the military planners want to accomplish. In fact, isolation of the perpetrator-group is in itself not impartial but required. Military interveners need to be careful during planning to stop the violence and not to instill even more hatred, or there will never be reconciliation in the area of operations. The military must be even-handed, based on a functioning rule of engagement, stopping the perpetrator-groups, but not allowing for reciprocation from the protected victims.

The final distinction in MARO is the escalatory dynamic. Mass atrocities can increase in intensity very quickly once they start. The emotional effects of hatred and fear can quickly unleash a killing frenzy. Perpetrator-groups may even speed up the killing once it has begun in order to complete the cleansing before intervention occurs.[50] Therefore, the U.S. military must prepare for quick intervention to stop the killing before it gains traction in the area.

---

[47] Sarah Sewall, Dwight Raymond, and Sally Chin, *Mass Atrocity Response Operations: A Military Planning Handbook*, 10.
[48] Ibid., 27-28
[49] Ibid., 26.
[50] Ibid., 23.

The MARO handbook also provides several operational implications planners must understand in order to negate mass atrocities. Of these, there are several pertinent implications air power can play an integral part. The first of these is intelligence.[51] Due to the inherent escalatory dynamics in genocide, timely information is the key to a successful MARO. Intelligence provides planners with the information to base decisions upon, both at the strategic and operational levels of war. Strategically, it can provide planners with the indicators of a likely mass atrocity. Once indications predict a mass atrocity will occur, the military must quickly align collection assets to provide planners with the scope of the mass atrocity. This information is critical to allow the National Command Authority to determine the response, if any. Operationally, intelligence provides the information required to base a myriad of decisions. Included in these is the disposition of enemy forces, the amount of force required to defeat the perpetrators of the mass atrocity, and which course of action is the best.

The next operational implication the MARO handbook provides is speed versus mass.[52] The last version of Army FM 3-0 stipulated that, "massing in a stability or civil support operation includes providing the proper forces at the right time and place to alleviate suffering and provide security."[53] Due to the potential for quick escalation during mass atrocities, this concept is even more critical. The handbook states, "In a conventional conflict, if an intervention arrives late, certain aspects can be "undone" – territory may be recaptured or prisoners released. In a MARO situation, the perpetrator has achieved success if the civilians it wishes to have killed are killed; no subsequent victory against the perpetrators will undo the civilian deaths."[54] Thus, the speed of response to MARO operations is crucial.

---

[51] Ibid., 29.
[52] Ibid.,33-34.
[53] U.S. Department of the Army, *Operations*, Field Manual 3-0 (Washington, DC: U.S. Department of the Army, February 2008), A-2.
[54] Sarah Sewall, Dwight Raymond, and Sally Chin, *Mass Atrocity Response Operations: A Military Planning Handbook*, 33.

The next operational implication provided in the handbook is the power of the witness.[55] As already eluded to in mental and moral isolation of the perpetrator-group, individuals are much less willing to commit acts of genocide when there is transparency to their operations. When looking at the perpetrator-group as a whole, transparency can undermine the genocidal acts by revealing their actions to the international community. Similarly, transparency exposes individuals to the scrutiny of the social environment.

The final operational implication of combating mass atrocity is the immediate need for non-military requirements.[56] The needs of the victims, and possibly the entire area of operation's population during a MARO could be substantial. Planners must therefore be prepared to conduct many of the same missions seen in COIN operations. These include civil security, provision for basic economic needs, restoration of essential services such as water, electricity, etc., and governance.[57] This implication definitely increases the military footprint in the area, and may not be conducive to a particular operation. If the military's directive is just to stop the violence from occurring, then planners must accordingly devise means to avoid these effects interrupting operations.

Phasing

MARO operations can span a wide spectrum of military actions, from prevention or deterrence, to intervention. Military planners must be adept at providing military options that span the numerous possibilities of a mass atrocity response. A flexible model to arrange MARO operations is the Phasing Model provided in Joint Publication 5-0. Not only is it consistent with MARO operations, but it also provides continuity to military planners, as it is the predominantly understood model.

---

[55] Ibid., 35-36.
[56] Ibid., 37-38.
[57] U.S. Department of the Army, *Counterinsurgency*, Field Manual 3-24 (Washington, DC: U.S. Department of the Army, December 2006), 2-2.

21

Phase 0 operations include the normal operations of the military, the steady state of affairs.[58] For the purposes of this paper, Phase 0 is ongoing before a future perpetrator-group plans an act of genocide. However, there may be indicators of a possible genocide. MARO operations can appear unrewarding during this phase, but are ultimately the greatest means to prevent or stop a mass atrocity. MARO during this phase should develop and arrange military security cooperation with regional partners, building the legitimacy of possible security forces. This collaboration alone could dissuade perpetrator-groups from planning genocide. It also would provide for partners to gather, share, and understand intelligence. Phase 0 operations should also include the appropriate escalation of ISR in locations with indicators of a possible genocide. Due to the escalatory dynamic of genocide, the appropriate surveillance of these hot spots could provide the time needed to react. To this end, planners must also have off-the-shelf plans established for genocide prone countries to quickly respond to deteriorating situations.

This paper's assumption is the transition from Phase 0 to Phase 1 occurs when there is evidence of impending genocide or initial acts of genocide have already occurred. Phase 1 operations include those missions with the intent of deterring the perpetrator-group from starting or continuing genocide. In order to do this, planners must demonstrate military capability to stop the genocide and prove the resolve of a coalition to commit to such operations.[59] Ideally, the establishment of national policy should be quick due to the possible escalation of killing. Thus, planners need to take the off-the-shelf plans and provide the National Command Authority (NCA) with a considerable list of options the United States could take. By doing this, the planners enable the NCA to commit to a MARO without necessarily prescribing the level of commitment. The faster the interveners can act, the better possibility of isolating the perpetrator-group and ending the genocide.

---

[58] U.S. Department of Defense, *Joint Operation Planning,* Joint Publication 5-0 (Washington, DC: U.S. Department of Defense, August 2011), III-42.

[59] Ibid.

These off-the-shelf plans should prescribe the deployment of forces in a deterrent posture, establish the command and control infrastructure, increase security cooperation and collaboration in the region, accomplish shows of force, and begin to collect mission-oriented intelligence. There is also an opportunity for planners to take preliminary steps to isolate the enemy, both mentally and morally. The use of indemnifying intelligence in the international media should prove the steps the perpetrator-group has taken toward genocide. This will morally isolate the enemy system from international support, spurring trade and arms embargoes. Planners should also attempt to mentally isolate the enemy by beginning the local narrative. The local narrative should include measures to explain the consequences of genocide as well as demonstrate the lack of impunity for individuals within the perpetrator-group.

The transition from Phase 1 to Phase II occurs once the degree of genocide has exceeded the level prescribed in national strategy. This degree is abstract in this paper, but relates to the objectives of the MARO. Ultimately the start of Phase II is dependent on the amount of effort the United States or the coalition is willing to commit to the operation. Thus, the transition from Phase 1 to Phase 2 could be transparent if the level of commitment is high, or not occur if the level of commitment is low. Phase 2 operations commence with the objective to seize the initiative from the perpetrator-group and begin to stop the genocide.[60]

According to JPUB 5-0, Phase 2 operations attempt to force an adversary to culminate offensively by halting aggression and denying their objectives.[61] In MARO, the perpetrator-group's objective is the mass killing of the victims. Therefore, planners are subject to a short timeline before the perpetrator-group achieves their objective, an un-reversible objective by any amount of force once achieved. Thus, the characterization of MARO in Phase 2 is a build-up of forces that enter the fight piecemeal. Initial efforts in this phase are two-fold: set the conditions for Phase 3 forces to enter the AO, and disrupt the capability for perpetrator-groups to commit

---

[60] Ibid., III-42-43
[61] Ibid.

genocide. Planners must aggressively use simultaneity and depth in order to exploit the enemy with forces as they filter into theater. To do this, planners need to isolate the perpetrator-group physically, mentally, and morally.

In Phase 2 operations, physical isolation of the enemy commences in earnest. The first step to physically isolate the enemy is to establish friendly freedom of movement. Notably, the primary effect to accomplish this is to enforce a no-fly zone and control the borders. Once established, planners are able to leverage the mobility of US military forces to destroy command and control networks. Planners can accomplish this by destroying leadership nodes within the enemy system or the communication architecture of it. Disruption of logistics will further isolate the enemy from the capability to commit genocide. Finally, military forces need to isolate the enemy from the victims by establishing safe havens or refugee camps.

Mental isolation will continue to develop in this phase as the local narrative becomes more influential. Precision targeting of communication from leadership and control nodes will cause disorder and confusion. Planners then need to supplant this lack of information with the local narrative focusing on retribution and justice. They can also coerce individuals within the perpetrator-group by replacing peer pressure with social pressure. Similarly, moral isolation will continue to develop during Phase 2 operations. Efforts to provide genocide evidence to the international community should be paramount. This will provide the momentum to eliminate military assistance to the perpetrator-group, but also foster relationships within the coalition to provide forces for possible Phase 4 and Phase 5 operations.

The dominate phase, Phase 3, focuses on breaking the enemy's will and achieving the objectives of the operation.[62] In traditional conflicts, planners can more clearly delineate between Phase 2 and Phase 3. However, MARO focuses on speed versus overwhelming mass, and the transition between the two phases is blurred. Planners must develop solutions in this phase to provide an overwhelming joint force capability at the critical time and place to destroy the efforts

---

[62] Ibid., III-43

24

of the perpetrator-group.[63] Isolation of the enemy is still occurring at every level but the predominance of effort lies in physical isolation. As planners sequence forces into the AO, they focus their efforts on the pockets of perpetrator-groups willing to continue the acts of genocide. The interveners must capture any remaining command and control networks. The military should control critical infrastructure canalizing the enemy and denying enemy logistics, effectively ending the economic means needed to continue genocide. This will also allow for precision targeting of the actual perpetrators, by either destroying or capturing them.

As Phase 3 achieves the strategic objective to end the mass killing, Phase 4 and 5 focuses on restoring order and transitioning to a sustained peace. Joint publication 5-0 identifies these two phases as the stabilize phase and enable civil authority phase respectively.[64] MARO in these phases differs in the prosecution of war criminals and security sector reform when compared to other operations, but many of the tasks are similar to operations of counter-insurgency and foreign internal defense. Thus, this paper will not focus on these two phases. The author would be remiss not to remind military practitioners that planning and operations in these phases are very difficult. Planners should review lessons learned from Operations Iraqi Freedom and Enduring Freedom while planning for these two phases.

Air Power Solutions

The paper thus far focused on a general view of genocide prevention and response operations using Boyd's work, the MARO handbook, and Joint Publications. However, the purpose of this paper is to reveal practical air power solutions to executing MARO. Airpower's speed, range, flexibility, precision, and lethality provide a spectrum of employment options perfectly suited for MARO.[65] Not only can airpower apply force against many facets of enemy

---

[63] Ibid.

[64] Ibid., III –43-44.

[65] U.S. Department of the Air Force, *Air Force Basic Doctrine, Organization, and Command*, Air Force Doctrine Document 1, (Washington, DC: U.S. Department of the Air Force, October 2011), 16.

25

power, but it is also less culturally intrusive.[66] Thus, the following section will delve into the core functions of air power to navigate through the opportunities air power can exploit to combat mass atrocities. These core functions break down the operational capabilities of the US Air Force into manageable areas to focus upon. While not directly parallel to the Joint Publication Phasing Model, the author will sort the core functions chronologically to the extent possible for MARO.

Additionally, this section of the paper will focus on those capabilities that play a specific and unique role in MARO, and are not encompassing of all of the different missions the US Air Force would need to execute in a proper MARO. The paper will then highlight the effects these functions could have in the 1994 Rwandan genocide. This paper does not intend to argue whether or not the United States should have reacted to the crisis in Rwanda. However, it presumes that in today's R2P environment there would have been a strong impetus to react from the international community. Thus, it will utilize the Rwandan genocide as a hypothetical case study, placing the events that occurred in 1994 in today's current technological and political environment.

Global Integrated ISR

The first core function the Air Force provides is Global Integrated Intelligence, Surveillance, and Reconnaissance (ISR). Global Integrated ISR is "the synchronization and integration of the planning and operations of sensors, assets, and processing, exploitation, dissemination systems across the globe."[67] While the Air Force plays a critical role in the collection and dissemination of intelligence, it is just a piece of the United States intelligence network. However, the Air Force intelligence community does provide some critical capabilities perfectly suited to MARO: global reach, persistence, and timeliness.

---

[66] Ibid., 19
[67] Ibid., 48.

26

The mobility of air assets provides on-demand reconnaissance throughout the area of operations readily accessible to commanders.[68] Limited national level intelligence sources are extremely over-tasked. The Air Force fills the gap for operational level intelligence, which is paramount when combating mass atrocities. The persistence of manned and more recently acquired Remotely Piloted Aircraft (RPA) denies enemies the ability to "hide" from collection efforts. Air Force ISR assets focus on detailed collection against target sets for an extended period of time.[69] These assets include satellites, reconnaissance aircraft, RPA, and electronic combat aircraft. These systems are perfectly suited to gather required intelligence as they can quickly deploy in a time-constrained environment, and have a minimal footprint on the area of operations.

The synergistic effects of USAF ISR assets with national means play a crucial role in Phase 0 and 1 operations. The constant feed of intelligence is critical to identify possible indicators of genocide. Not only do they provide the impetus for action, but they can also isolate the enemy. MQ-9 and MQ-1 RPA utilize full motion video to stream EO/IR imagery or TV to ground control stations worldwide.[70] The ability to record video evidence of perpetrators committing genocide could be strategically instrumental to hamper enemy operations. The video evidence of perpetrators in the act of genocide could prove violations of the UN Convention on the Prevention and Punishment of the Crime of Genocide. Once the international community is convinced of these violations, they are more apt and willing to act on it. This would morally isolate the perpetrator group and place international pressure to end it, either economically or militarily or both.

ISR assets are also critical to identify critical nodes and linkages within the enemy system. RC-135V/W aircraft can "detect, identify and geo-locate signals throughout the

---

[68] U.S. Department of the Air Force, *Operations and Organization*, Air Force Doctrine Document 2, (Washington, DC: U.S. Department of the Air Force, October 2011), 7.
[69] Ibid.
[70] MQ-1B Factsheet, http://www.af.mil/information/factsheets/factsheet.asp?id=122 (accessed November 29, 2011).

electromagnetic spectrum. The mission crew can then forward gathered information in a variety of formats to a wide range of consumers via Rivet Joint's extensive communications suite."[71] The ability to locate critical nodes, either leadership or communication nodes, is critical intelligence planners should utilize. This information in the hands of aviators conducting air interdiction or strategic attack could physically remove the node, or alternatively, this information could exploit the enemy system.

If the United States decided to react to an impending genocide in Rwanda, actionable intelligence would have been required at the strategic and operational level of war. The evolution of RPA technology would have been a critical factor to deliver just this. The persistence over the battlefield of ISR and RPAs would have dramatically increased the level of intelligence available. General Dallaire, head of the United Nations forces located within Rwanda repeatedly asked for reinforcements before and following the April 6th shoot-down of the plane transporting Rwandan President Habyarimana back in country.[72] This event was the spark in the already volatile country, leading to Hutus systematic killing of Tutsis and their supporters. In today's environment, this would have been the final indicator of the impending genocide. Rapid deployment and employment of intelligence collection efforts would have provided several benefits.

The primary benefit ISR could have played in the wake of the plane crash would have been the ability to document acts of genocide. Specifically, RPAs could provide video evidence of multiple acts of genocide, taking the possible threat of genocide into reality. This evidence would have most likely provided the United Nations and President Clinton the international support to react to the genocide. The earlier it was identified, the sooner a coalition could have been built to stop it. The next benefit of ISR in the Rwanda scenario would have been the quick

---

[71] RC-135V/W Factsheet, http://www.af.mil/information/factsheets/factsheet.asp?id=121 (accessed November 29, 2011).
[72] Romeo A. Dallaire, *Shake Hands with the Devil: The Failure of Humanity in Rwanda* (New York, NY: Carroll & Graf Publishers, 2003), 84.

identification of the perpetrator group and its leadership. Specifically, early identification of the Interahamwe militia and leadership would be crucial for physical and mental isolation. Identifying the communication nodes from the Interahamwe would not have been too difficult as much of the hate propaganda stemmed from the RTLM radio network formed specifically by the Interahamwe for this function.[73] Once identified, the coalition would have multiple options to isolate the network and perpetrators, the Hutu extremists.

Special Operations

Special operations are operations conducted in politically sensitive environments to achieve an objective in any national instrument of power: military, diplomatic, informational, or economic.[74] While the predominant percentage of forces lie in other elements of the Department of Defense, Air Force elements are a force multiplier and play an even more critical role in MARO situations. One subset of the Special Operations core functions that specifically plays a crucial role in MARO is Information Operations.[75]

Information Operations (IO) is the integrated employment of electronic warfare and influence operations to influence, disrupt, or corrupt human decision-making capability. If viewed through Boyd's lens, these are the primary ways to mentally isolate the enemy in order to achieve one's objective.[76] Planners must utilize the detailed intelligence collection in order to prioritize targets within the enemy system to systematically and continually disorient the enemy.

The first primary method to disorient the individuals within the perpetrator-group is to cut the communication linkage in the enemy system. The US Air Force's EC-130H is a tactical

---

[73] Yanagizawa-Drott, David. *Propaganda and Conflict: Theory and Evidence from the Rwandan Genocide,* http://www.hks.harvard.edu/fs/dyanagi/Research/RwandaDYD.pdf (accessed March 2, 2012), 2.

[74] U.S. Department of Defense, *Department of Defense Dictionary of Military and Associated Terms,* Joint Publication 1-02, (Washington, DC: U.S. Department of Defense, October 2011), 266.

[75] U.S. Department of the Air Force, *Air Force Basic Doctrine, Organization, and Command,* Air Force Doctrine Document 1, 50.

[76] U.S. Department of the Air Force, *Information Operations,* Air Force Doctrine Document 3-13, (Washington, DC: U.S. Department of the Air Force, January 2005), 12.

weapon system designed to disrupt enemy command and control communications.[77] By jamming communications of the perpetrator-group, the EC-130H would limit the leadership from communicating with its subordinates, rendering them to more primitive means of communications, which takes more time. In the interim, the EC-130H can employ offensive counter-information to further disorient the enemy and mentally isolate them.[78]

As mentioned earlier, MQ-9 and MQ-1 RPA utilizing full motion video to stream EO/IR imagery or TV to commanders could prove critical.[79] Video evidence of perpetrators committing genocide could prove operationally useful as well, to garner support of the local populace. If planners decide to broadcast this evidence on local TV stations, the general populace faces the harsh reality of the genocide. Planners must carefully navigate the video's effects as it could spurn retribution, but it could also prove useful in gaining support of those opposed or initially indifferent to the genocide. Local religious and political leaders may unite against the perpetrator-group further isolating them from their support base. Social pressure from the local community or even family members subjects individuals within the perpetrator-group to scrutiny from other members in their community.

The US Air Force provides a niche air asset capable of broadcasting this video evidence or portion of the local narrative. The EC-130J aircraft conducts information operations by broadcasting on AM, FM, HF, TV or military communication bands.[80] If planners carefully synchronize the effects of the EC-130H and EC-130J, they can effectively cut off internal electromagnetic communication and replace it with their own local narrative, such as the video evidence of the genocide.

---

[77] EC-130H Factsheet, http://www.af.mil/information/factsheets/factsheet.asp?id=190 (accessed November 30, 2011).
[78] Ibid.
[79] MQ-1B Factsheet, http://www.af.mil/information/factsheets/factsheet.asp?id=122 (accessed November 29, 2011).
[80] EC-130J Factsheet, http://www.af.mil/information/factsheets/factsheet.asp?id=182 (accessed December 1, 2011).

Individuals are less willing to commit genocide if they are accountable for their actions. Simply filming the individuals may stop a significant portion of the perpetrator-group. However, as planners develop the local narrative, combining this evidence with the actual capture, trial, and punishment of individuals would further mentally isolate the individuals. The final piece this evidence provides is the actual interdiction of the individual by targeting them with precision attack systems. If no other effort provides the motivation for individuals to stop committing genocide, fear of retribution and death could work.

The development of a local narrative would have been critical in Rwanda as the genocide expanded exponentially after April 6th, 1994. One of the first targets of U.S. Air Force special operations would have been to eliminate the radio network, RTLM. This network was formed by the Interahamwe to foster hate crimes against Tutsis and Hutu moderates.[81] This station continuously broadcast hate propaganda and at times transmitted the location of Tutsis fleeing from perpetrators. The EC-130H radio jamming capability could limit communication in the initial stages of the genocide. Similarly, the RC-135's ability to detect and locate those stations would provide the perfect target for strikes having the most impact.

Rwanda's dependence on the radio to transmit information would also prove very useful as U.S. forces developed a local narrative. The militia forces conducted the first acts of genocide, but soon this evolved into the general Hutu populace conducting the massacres. Thus, the narrative would need to focus on slowing the genocide by influencing the Rwandan populace, both Hutu and Tutsi from stopping the killings. The EC-130J broadcast capability would use the now open radio network for this. This narrative could be used to positively influence the people but also demonstrate there would be punishment for those who committed genocide.

---

[81] Yanagizawa-Drott, David. *Propaganda and Conflict: Theory and Evidence from the Rwandan Genocide*, 3.

Air Superiority

Air superiority is the degree of air dominance that permits the conduct of operations in land, sea, and air without interference from another force.[82] In most MARO situations, air superiority would be one of the first priorities of the US Air Force. Once the US achieves a degree of air dominance, subsequent air and ground operations could take place to include ISR, Air Interdiction, and Airlift. Thus, the US Air Force would employ a wide variety of fighter aircraft, tanker aircraft, and electronic attack aircraft to enforce a no-fly zone over the area of operations. The effects of this core function would be two-fold in a MARO situation.

The first effect resulting from a no-fly zone would be a degree of physical isolation of the perpetrator-group. This isolation would disable freedom of movement for the enemy, as it would limit the use of the air domain for transportation and logistics. The no-fly zone would also limit the possible effects of genocide, as perpetrator-groups would be unable to use attack aircraft and helicopters. While a no-fly zone is not sufficient to end ground forces from committing genocide, it would end the use of enemy aviation to commit genocide critically slowing the rate of genocide.

The second effect of air superiority is opening the air domain to friendly and neutral parties. Freedom of movement within the air domain is a force multiplier for multiple reasons. The Department of State and non-governmental agencies would have freedom of access to conduct humanitarian and nation-building missions, while minimizing their exposure to possible retaliatory forces from the perpetrator-group. The opening of the air domain allows for subsequent air operations crucial to the success of a MARO, as US Air Force assets would be unhindered.

---

[82] U.S. Department of the Air Force, *Air Force Basic Doctrine, Organization, and Command*, Air Force Doctrine Document 1, 45.

Global Precision Attack

Global Precision Attack encompasses the range of air operations designed to strike rapidly any target to provide effects needed by commanders.[83] These effects can be both kinetic, taking the form of bomb dropping, to non-kinetic, such as shows of force. This core function is obviously the most lethal but can also provide a level of deterrence to the perpetrator-group. Thus, planners should carefully balance the use of kinetic versus non-kinetic operations within this sphere, in order to minimize collateral damage and civilian casualties and maximize the isolation of the enemy. Global Precision Attack isolates the enemy primarily physically and mentally.

Once air superiority is established, the US Air Force is able to bring the full arsenal of its capabilities to bear against the perpetrator group. In Phase 1 and 2 operations, air interdiction could be quite fruitful in isolating the enemy. Air interdiction encompasses missions conducted to disrupt, divert, delay, or destroy an enemy's ability to bear effectively against friendly forces or other objectives, namely the victims in a genocide scenario.[84] Fighter and bomber aircraft could target any node or linkage that would disorient and hamper the survivability of the perpetrator-group.

Planners must again carefully assess any targets for second and third level effects, but the primary consideration in MARO is the survival of the victims. Air superiority restricted freedom of movement in the air domain, as air interdiction can do on the ground. Critical road networks and bridges as targets would severely hamper the transportation network of the perpetrator-group. This would not only limit face-to-face communication, but also isolate pockets of perpetrators cutting them off from logistics required to continue conducting the mass atrocity. Attacking the transportation network could also establish safe havens for the victims by canalizing the enemy

---

[83] Ibid., 48-49.
[84] U.S. Department of the Air Force, *Counterland Operations*, Air Force Doctrine Document 3-03, (Washington, DC: U.S. Department of the Air Force, July 2011), viii.

33

away from them. Operational level intelligence is critical as the degree of planning is quite sophisticated.

As the MARO develops, missions designed to deter the individuals within the perpetrator-group can become more effective. Shows of force over enemy locations would mentally isolate individuals as the risk-reward dynamic begins to develop. The development of local narrative in a given MARO would only gain ground as perpetrators realize they are accountable for their actions. Direct targeting of perpetrators in either air interdiction or close air support missions would only further their fears. However, any bombs dropped near civilian populations needs to assess possible instances of collateral damage. The use of low collateral damage weapons could limit the effects, but as we have seen recently in Operation Enduring and Iraqi Freedom, mistakes could be more detrimental than beneficial.

A final piece to isolate the enemy is the direct targeting of leadership and key nodes within the perpetrator-group. These targets would be strategic vice operational, and so commanders may take more risk of collateral damage and detrimental effects. However, the payoff could prove quite useful in mentally isolating the perpetrator-group. Removal of the leadership from the equation forces the enemy system into disequilibrium. During the interval while it recovers, individuals within the perpetrator-group are more receptive to the local narrative and actions taken by US forces.

The genocide in Rwanda, as with any genocide, would be difficult for kinetic options due to the inherent risk of collateral damage. However, several key targets identified in retrospect could have proved critical. The first would be the communication nodes of the Interahamwe, namely the RTLM network. Specific leadership within this militia was also available to be targeted. However, most of the perpetrator group was the general Hutu public. Thus, specific kinetic options would be limited. However, shows of force to disperse crowds would be useful and land forces responded to crisis area.

Rapid Global Mobility

The US Air Force is quite aware of the implications of strategic airlift and theater mobility. The rapid and flexible air mobility operations provide commanders the options required to achieve objectives in a myriad of scenarios. Airlift cannot only transport forces and logistics into theater, but provides for further on humanitarian support during later phases. If required, they are also available to physically isolate the enemy from the victims by evacuating the victims to safe havens or neighboring countries. While air mobility operations are not necessarily unique when considering MARO, the expeditionary nature provides the speed and flexibility required to diffuse mass atrocities before they escalate.[85] Thus, they deserved mention in this paper and could have proved quite useful prior to or just after the onset of genocide in Rwanda.

Recommendation

The analysis of this paper revealed practical air power solutions to the problem of MARO. While not all inclusive of the missions performed in a MARO by the US Air Force, the paper intended to identify those not currently planned or considered. Some of the missions proposed in this paper fall under the purview of training regimes of respective platforms already established, others do not. However, the US Air Force will not specify a need to add a training requirement until identified. Thus, there are three broad recommendations this paper provides.

First, combatant commanders need to develop specific MARO plans for their geographic command. These "off-the-shelf" plans should be complete with a myriad of options giving the National Command Authority the impetus to support the MARO. This advanced planning will be fruitful for two reasons. Most importantly, due to the escalatory dynamic of MARO, the quickest response possible will prevent further death. The more substantial the "off-the-shelf" plan is, the faster the U.S. military can respond. These plans will minimize the delay of moving forces forward in support of the MARO. Fully developed plans will also specify the overall conduct and

---

[85] U.S. Department of the Air Force, *Air Mobility Operations*, Air Force Doctrine Document 3-17, (Washington, DC: U.S. Department of the Air Force, July 2011), 13.

35

direction of the operation, identifying the forces and mission-sets required. These capabilities and mission-sets will drive the requirement for training at home.

Training is the key to successfully executing a MARO. Any lapses in training should be quickly determined and protocol developed to train airmen in their respective platforms. This training alone is not sufficient, as a MARO must be integrated and synchronized at all levels. Air Force unit exercises would integrate air power, focusing on those specific tasks required in a MARO. These training exercises, in conjunction with joint training would fully prepare units identified in a MARO plan.

The final recommendation for implementation is the inclusion of MARO into Air Force and Joint doctrine. Doctrine is a conceptual framework or a foundation for the military to plan. Not all MARO situations are going to be the same but doctrine highlights the skills required to execute the operation. It also compels airmen to fix deficiencies in training programs and capabilities in the Air Force.

Conclusion

The objective of this paper was to identify how the United States Air Force best executes its role in Mass Atrocity Response Operations and provide practical air power solutions to the problems it will face. It traced current literature as a starting point to understand the complexities of genocide and current national policy regarding it. It is evident, should the trend continue, that national leadership and momentum favor prevention and intervention to the global problem. Recent developments provide even stronger measures for the prevention of genocide. On August 11, 2011, President Obama issued a Presidential Directive on Mass Atrocities, Presidential Study Directive-10.[86] This directive created an interagency Atrocities Prevention Board to develop strategies to prevent mass atrocities and implement a stronger coordination between the United

---

[86] Barack H. Obama, *Presidential Directive on Mass Atrocities* (Washington, DC: The White House, Presidential Study Directive 10, August 2011), 1.

States and its allies. This momentum towards prevention differs from United States history, as it seems intervention is both in the nation's interest and within its values.

To develop the thesis, the paper used the work of Colonel John Boyd and his work on isolation of the enemy. It built upon his foundation to identify critical nodes and linkages in the perpetrator-group and pointed out several key areas the military could exploit. By using Joint Publication and the PKSOI MARO project, it illustrated the current framework of operations and the implications to the military if it undertook this endeavor. The paper synthesized the work of Boyd, the MARO project, and Joint Publication to identify the niche that the Air Force could fulfill. Finally, by reviewing Air Force Doctrine, it laid out practical air solutions to the problem of genocide intervention.

The 1994 genocide in Rwanda was one of many ink spots on humanity. The horrific events led to 800,000 people dead in just over three months.[87] Although the goal of this paper was not to advocate for a United States responsibility to protect victims of genocide, it is hard to separate national interest from idealistic values. What if the United States military had intervened? The outcome would most likely have been remarkably different. As national leaders move forward with a policy of prevention and intervention, the United States military and Air Force must adapt and prepare to defeat humanity's enemy.

---

[87] Romeo A. Dallaire, *Shake Hands with the Devil: The Failure of Humanity in Rwanda,* xvii.

BIBLIOGRAPHY

Albright, Madeleine, and Cohen, William, *Preventing Genocide: A Blueprint for U.S. Policymakers.* Washington, DC: Genocide Prevention Task Force, United States Holocaust Memorial Museum, 2008.

Annan, Kofi A. *The Millennium Report of the United Nations Secretary-General, 2000.* New York: United Nations, Department of Public Information, 2000.

_____. "Undersecretary-General of the United Nations." *The East African,* March 1996.
Bush, George W., *The National Security Strategy of the United States of America.* Washington, DC: The White House, March 2006.

Craig, Robert T. and Tracy, Karen, "Grounded Practical Theory: The Case of Intellectual Discussion." *Communication Theory 5,* issue 3 (March 1995): 248-272.

Dallaire, Lt Gen (Ret) Romeo. *Shake Hands with the Devil: The Failure of Humanity in Rwanda.* New York: Carroll & Graf, 2003.

EC-130H Factsheet, http://www.af.mil/information/factsheets/factsheet.asp?id=190 (accessed November 30, 2011).

EC-130J Factsheet, http://www.af.mil/information/factsheets/factsheet.asp?id=182 (accessed December 1, 2011).

Evans, Gareth. *The Responsibility to Protect: Ending Mass Atrocity Crimes Once and For All.* Washington, DC: Brookings Institution Press, 2008.

Evans, Gareth and Sahnoun, Mohamed. *The Responsibility to Protect.* Ottowa, CA: International Development Research Centre, 2001.

MQ-1B Factsheet, http://www.af.mil/information/factsheets/factsheet.asp?id=122 (accessed November 29, 2011).

Obama, Barack H., *The National Security Strategy of the United States of America.* Washington, DC: The White House, May 2010.

_____. *Presidential Directive on Mass Atrocities,* (Washington, DC: The White House, Presidential Study Directive 10, August 2011).

Osinga, Frans P.B. *Science, Strategy and War: The strategic theory of John Boyd.* New York: Routledge, 2007.

Peifer, Douglas., "Genocide and Air Power." *Strategic Studies Quarterly 2.2* (Summer 2008): 93-124.

Power, Samantha. *"A Problem from Hell:" America and the Age of Genocide.* New York: Basic Books, 2002.

RC-135V/W Factsheet, http://www.af.mil/information/factsheets/factsheet.asp?id=121 (accessed November 29, 2011).

Sewall, Sarah, and Raymond, Dwight, and Chin, Sally. *Mass Atrocity Response Operations: A Military Planning Handbook.* Charleston: Createspace, 2010.

U.S. Department of the Air Force, *Air Force Basic Doctrine, Organization, and Command,* Air Force Doctrine Document 1. Washington, DC: U.S. Department of the Air Force, October 2011.

_____. *Operations and Organization,* Air Force Doctrine Document 2. Washington, DC: U.S. Department of the Air Force, October 2011.

_____. *Counterland Operations,* Air Force Doctrine Document 3-03. Washington, DC: U.S. Department of the Air Force, July 2011.

_____. *Information Operations,* Air Force Doctrine Document 3-13. Washington, DC: U.S. Department of the Air Force, January 2005.

_____. *Air Mobility Operations,* Air Force Doctrine Document 3-17. Washington, DC: U.S. Department of the Air Force, July 2011.

U.S. Department of the Army, *Operations,* Field Manual 3-0. Washington, DC: U.S. Department of the Army, February 2008.

_____. *Counterinsurgency,* Field Manual 3-24. Washington, DC: U.S. Department of the Army, December 2006.

U.S. Department of Defense, *Joint Operation Planning,* Joint Publication 5-0. Washington, DC: U.S. Department of Defense, August 2011.

_____. *Department of Defense Dictionary of Military and Associated Terms,* Joint Publication 1-02. Washington, DC: U.S. Department of Defense, October 2011.

United Nations General Assembly. *2005 World Summit Outcome.* 60th sess., September 15, 2005.
_____. *Convention on the Prevention and Punishment of the Crime of Genocide,* Adopted by Resolution 260, December 9, 1948.

Waller, James. *Becoming Evil: How Ordinary People Commit Genocide and Mass Killing.* 2nd ed. Oxford: Oxford University Press, 2007.

Yanagizawa-Drott, David. *Propaganda and Conflict: Theory and Evidence from the Rwandan Genocide,* http://www.hks.harvard.edu/fs/dyanagi/Research/RwandaDYD.pdf (accessed March 2, 2012).

www.ingramcontent.com/pod-product-compliance
Lightning Source LLC
Chambersburg PA
CBHW080620180526
45168CB00007B/2997